IN SPACE

Earth

Popcorn

Chris Oxlade

WAYLAND

Explore the world with **Popcorn** - your complete first non-fiction library.

Look out for more titles in the **Popcorn** range. All books have the same format of simple text and vibrant images. Text is carefully matched to the pictures to help readers to identify and understand key vocabulary.
www.waylandbooks.co.uk/popcorn

First published in 2009 by Wayland
Copyright Wayland 2009

Wayland
338 Euston Road
London NW1 3BH

Wayland Australia
Level 17/207 Kent Street
Sydney NSW 2000

Editor: Julia Adams
Designer: Robert Walster
Picture Researcher: Julia Adams

British Library Cataloguing in Publication Data
Oxlade, Chris.
    Earth. -- (Popcorn. In space)
    1. Earth--Juvenile literature.
    I. Title II. Series
    525-dc22

ISBN 978 0 7502 5774 9

Printed and bound in China

Wayland is a division of Hachette Children's Books,
an Hachette UK Company

www.hachette.co.uk

Acknowledgements:
Alamy: Luiz C. Marigo 20, Reinhard Dirscherl 21; iStockphoto: 17; NASA: front cover; SOHO/EIT: 10; NASA/Goddard Space Flight Centre: 9; NASA/JPL-Caltech/T. Pyle (SSC): 18/19; Shutterstock: Pichugin Dmitry 1, 7, Stephen Aaron Rees 2, 4, Jonathan Feinstein 5, Anson Hung 2, 6, AridOcean 8, Armin Rose 12, Alessio Ponti 13, 15;
Tudor Photography: 22, 23
Illustrations: Graham Rich

Contents

🌍 Our Earth

The place were we live is called Earth.

It is made up of land and water.

The water forms oceans, lakes and rivers.

This is the Atlantic Ocean. Which sea or ocean is nearest to you?

The Earth's land forms continents. There are six continents. Do you know which continent you live on?

This is what the Earth's land looks like from the sky.

Earth's features

The Earth has many different landscapes. There are long chains of mountains called mountain ranges. There are wide flat plains and deep valleys and canyons.

The Grand Canyon in the USA is 1,900 metres deep.

Some land is covered with thick forests. Other areas are covered with ice and snow. In some places there is bare rock or sandy desert.

The Himalayas are the highest mountains in the world.

The Earth's shape

When you look at a map of the Earth, the Earth looks flat. You need to look at a photograph of the Earth from far away to see its shape.

The yellow areas on this map are mountains.

Hundreds of years ago, people thought the Earth was flat.

This is a photograph of the Earth from space. From here, the Earth looks like a big ball. This shape is called a sphere.

 Sunlight

The Sun is a huge, glowing ball of
fiery hot gas. It is the Earth's closest star.
The Sun gives the Earth light and warmth.

The Sun looks like a ball of fire.

Never look
straight at the
Sun! It can
harm your
eyes.

The Sun's rays warm the Earth. Some parts of the Earth are closer to the Sun. The rays of the Sun warm these parts up the most.

The hottest part of the Earth is the area around the middle.

Earth

Sun

Different regions

The Earth has different areas called regions. The polar regions are at the top and bottom of the Earth. They are very cold because the Sun's rays hardly reach them.

Penguins live in the Antarctic. It is the polar region at the bottom of the Earth.

The area around the middle of the Earth is the equatorial region. It is very hot here all year round.

There are many deserts in the equatorial region.

Day and night

The Earth spins around. This means that the Earth's regions all have day-time and night-time. It takes the Earth 24 hours to spin around once.

The time it takes the Earth to spin around once is called a day.

Earth

Sun

The Sun lights up one side of the Earth.
On this side, it is day-time. On the other
side, it is night-time.

The Sun is
shining on
this side of
the Earth.

Moving around the Sun

The Earth moves around the Sun.

It moves in a giant circle, called an orbit.

The Earth moves around the Sun once every 365 days.

The time it takes the Earth to move
around the Sun once is called a year.

A year is the time it takes
from one birthday to the next.

The solar system

The Earth is part of a group of planets. There are eight planets in this group. Some planets are smaller than the Earth. Some are much bigger.

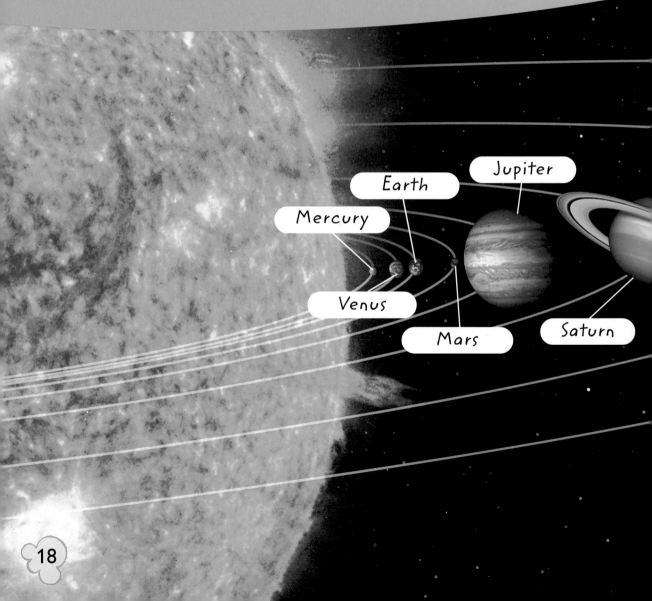

Mercury

Venus

Earth

Mars

Jupiter

Saturn

Together the planets and the Sun make up the solar system. The Sun is at the centre of the solar system.

Uranus

Neptune

Earth is the third planet from the Sun.

Life on Earth

The Earth is the only planet in the solar system with life. Animals and plants live on the land, in the oceans and in the air.

Thousands of different animals make their homes in rainforests.

Animals and plants cannot live without water. The Earth is the only planet with water in the solar system.

Many animals live in the deep oceans.

The Earth is a very special place. We must take care of it.

Make a day and night clock

Make a day and night clock to help you practise telling the time.

You will need:
- an empty cereal box
- paintbrush and paints
- 2 paper plates
- 2 paper fasteners
- cardboard • pencil
- coloured pens
- glue • scissors

1. Paint one side of a cereal box yellow and the other side black.

2. Glue one plate to each side of the box. Draw a clock face onto each plate.

3. Draw two long hands and two short hands onto the cardboard. Cut them out.

4. Use the paper fasteners to attach a long hand and a short hand to each clock face.

5. Paint the Moon and stars on the night-time side.

6. Paint the Sun and flowers onto the day-time clock.

Glossary

continent one of the six enormous pieces of land on the Earth

desert a place where it is very dry and there are few plants

map a drawing that shows where things are on the land

plain a very large flat area of land

ray light or heat travelling in a straight line

region an area of the Earth

Index